Copyright © 2011 XAMonline, Inc.
All rights reserved. No part of the material protected by this copyright notice may be reproduced or utilized in any form or by any means, electronic or mechanical, including photocopying, recording or by any information storage and retrievable system, without written permission from the copyright holder.

To obtain permission(s) to use the material from this work for any purpose including workshops or seminars, please submit a written request to:

XAMonline, Inc.
25 First Street, Suite 106
Cambridge, MA 02141
Toll Free: 1-800-509-4128
Email: info@xamonline.com
Web: www.xamonline.com
Fax: 1-617-583-5552

Library of Congress Cataloging-in-Publication Data

Wynne, Sharon A.
    PRAXIS II Middle School Mathematics 0069 Practice Test 2: Teacher Certification / Sharon A. Wynne. -1st ed.
    ISBN: 978-1-60787-124-8
    1. PRAXIS II Middle School Mathematics 0069 Practice Test 2.   2. Study Guides
    3. PRAXIS   4. Teachers' Certification & Licensure   5. Careers

Disclaimer:
The opinions expressed in this publication are the sole works of XAMonline and were created independently from the National Education Association, Educational Testing Service, or any State Department of Education, National Evaluation Systems or other testing affiliates.

Between the time of publication and printing, state specific standards as well as testing formats and website information may change that is not included in part or in whole within this product. Sample test questions are developed by XAMonline and reflect similar content as on real tests; however, they are not former tests. XAMonline assembles content that aligns with state standards but makes no claims nor guarantees teacher candidates a passing score. Numerical scores are determined by testing companies such as NES or ETS and then are compared with individual state standards. A passing score varies from state to state.

Printed in the United States of America   œ-1
PRAXIS II Middle School Mathematics 0069 Practice Test 2
ISBN: 978-1-60787-124-8

Praxis Middle School Mathematics 0069
Post-Test Sample Questions

DIRECTIONS: Read each item and select the best response.

1. $7t - 4 \cdot 2t + 3t \cdot 4 \div 2 =$
   (Average) (Skill 1.2)

   A. 5t

   B. 0

   C. 31t

   D. 18t

2. Which statement is an example of the identity axiom of addition?
   (Easy) (Skill 1.3)

   A. $3 + -3 = 0$

   B. $3x = 3x + 0$

   C. $3 \cdot \frac{1}{3} = 1$

   D. $3 + 2x = 2x + 3$

3. Which of the following does not correctly relate an inverse operation?
   (Average) (Skill 1.4)

   A. $a - b = a + -b$

   B. $a \times b = b \div a$

   C. $\sqrt{a^2} = a$

   D. $a \times \frac{1}{a} = 1$

4. Change $.\overline{63}$ into a fraction in simplest form.
   (Average) (Skill 1.6)

   A. 63/100

   B. 7/11

   C. 6 3/10

   D. 2/3

5. Which of the following is an irrational number?
   (Easy) (Skill 1.7)

   A. .362626262...

   B. $4\frac{1}{3}$

   C. $\sqrt{5}$

   D. $-\sqrt{16}$

6. $(3.8 \times 10^{17}) \times (.5 \times 10^{-12})$
   (Average) (Skill 1.8)

   A. 19×10^5

   B. 1.9×10^5

   C. 1.9×10^6

   D. 1.9×10^7

7. Given even numbers x and y, which could be the LCM of x and y?
   (Average) (Skill 1.11)

   A. $\frac{xy}{2}$

   B. 2xy

   C. 4xy

   D. xy

8. Simplify: $\frac{10}{1+3i}$
   (Rigorous) (Skill 1.16)

   A. $-1.25(1-3i)$

   B. $1.25(1+3i)$

   C. $1+3i$

   D. $1-3i$

9. Evaluate $3^{1/2}(9^{1/3})$
   (Rigorous) (Skill 1.17)

   A. $27^{5/6}$

   B. $9^{7/12}$

   C. $3^{5/6}$

   D. $3^{6/7}$

10. Simplify: $\sqrt{27}+\sqrt{75}$
    (Rigorous) (Skill 1.17)

    A. $8\sqrt{3}$

    B. 34

    C. $34\sqrt{3}$

    D. $15\sqrt{3}$

11. $\frac{3.5 \times 10^{-10}}{0.7 \times 10^{4}}$
    (Rigorous) (Skill 1.17)

    A. 0.5×10^{6}

    B. 5.0×10^{-6}

    C. 5.0×10^{-14}

    D. 0.5×10^{-14}

12. Solve for x and y:
    x = 3y + 7
    7x + 5y = 23
    (Rigorous) (Skill 1.19)

    A. (−1, 4)

    B. (4, −1)

    C. ($\frac{-29}{7}$, $\frac{-26}{7}$)

    D. (10, 1)

13. Which of the following is a factor of $k^3 - m^3$?
    (Average) (Skill 1.20)

    A. $k^2 + m^2$

    B. $k + m$

    C. $k^2 - m^2$

    D. $k - m$

14. Which of the following is a term in an expansion of the binomial $(x+y)^7$?
    (Rigorous) (Skill 1.20)

    A. $7x^7$

    B. $21x^5y^2$

    C. $21x^4y^3$

    D. $7xy^7$

15. Solve for x: $18 = 4 + |2x|$
    (Average) (Skill 1.21)

    A. $\{-11, 7\}$

    B. $\{-7, 0, 7\}$

    C. $\{-7, 7\}$

    D. $\{-11, 11\}$

16. Determine the area of the shaded region of the trapezoid in terms of x and y.
    (Rigorous) (Skill 2.2)

    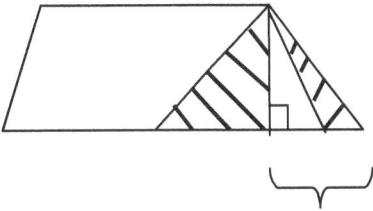

    A. $4xy$

    B. $2xy$

    C. $3x^2y$

    D. There is not enough information given

17. Find the length of a box with surface area of 94 sq. ft. with a width of 3 feet and a depth of 4 feet.
    (Rigorous) (Skill 2.2)

    A. 3 ft.

    B. 4 ft.

    C. 5 ft

    D. 6 ft.

18. Ginny and Nick head back to their respective colleges after being home for the weekend. They leave their house at the same time and drive for 4 hours. Ginny drives due south at the average rate of 60 miles per hour, and Nick drives due east at the average rate of 60 miles per hour. What is the straight-line distance between them, in miles, at the end of the 4 hours?
    (Rigorous) (Skill 2.3)

    A. $120\sqrt{2}$

    B. 240

    C. $240\sqrt{2}$

    D. 288

19. When you begin by assuming the conclusion of a theorem is false, then show that through a sequence of logically correct steps you contradict an accepted fact, this is known as
    (Easy) (Skill 2.6)

    A. Inductive reasoning

    B. Direct proof

    C. Indirect proof

    D. Exhaustive proof

20. Given $l_1 \parallel l_2$ which of the following is true?
    (Average) (Skill 2.6)

    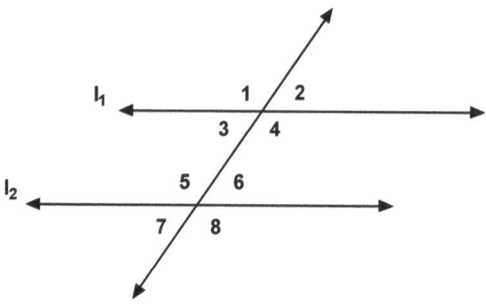

    A. ∠1 and ∠8 are congruent and alternate interior angles

    B. ∠2 and ∠3 are congruent and corresponding angles

    C. ∠3 and ∠4 are adjacent and supplementary angles

    D. ∠3 and ∠5 are adjacent and supplementary angles

21. In the figure above, what is the value of x?
    (Rigorous) (Skill 2.7)

    A. 50

    B. 60

    C. 75

    D. 80

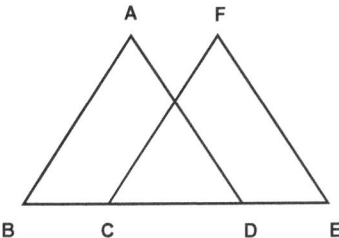

22. Which postulate could be used to prove △ABD ≅ △CEF, given BC ≅ DE, ∠C ≅ ∠D, and AD ≅ CF?
    (Average) (Skill 2.7)

    A. ASA

    B. SAS

    C. SAA

    D. SSS

23. Given that QO⊥NP and QO=NP, quadrilateral NOPQ can most accurately be described as a
    (Easy) (Skill 2.8)

    A. Parallelogram

    B. Rectangle

    C. Square

    D. Rhombus

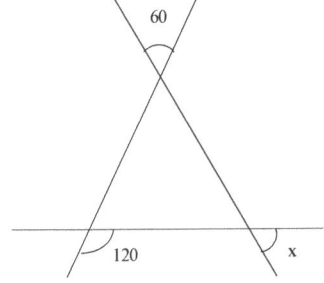

Note: Figure not drawn to scale.

24. What is the slope of any line parallel to the line
    2x + 4y = 4?
    (Rigorous) (Skill 2.12)

    A. -2

    B. -1

    C. -½

    D. 2

25. Find the midpoint of (2, 5) and (7, –4).
    (Average) (Skill 2.12)

    A. (9, –1)

    B. (5, 9)

    C. (9/2, –1/2)

    D. (9/2, 1/2)

26. Which set illustrates a function?
    (Easy)(Skill 3.2)

    A. {(0,1) (0,2) (0,3) (0,4)}

    B. {(3, 9) (–3, 9) (4,16) (– 4,16)}

    C. {(1, 2) (2, 3) (3, 4) (1, 4)}

    D. { (2,4) (3,6) (4,8) (4,16) }

27. Graph the solution:
    $|x| + 7 < 13$
    (Rigorous) (Skill 3.3)

    A)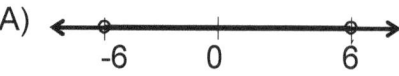

    B)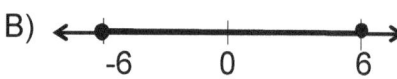

    C)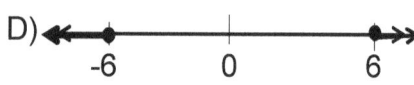

    D)

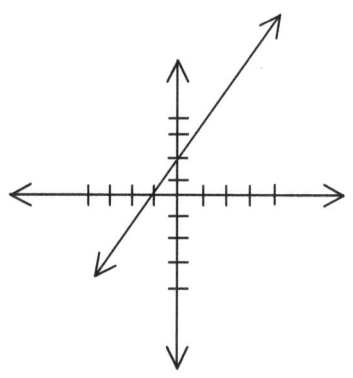

28. What is the equation of the above graph?
    (Average) (Skill 3.3)

    A. $2x + y = 2$

    B. $2x - y = -2$

    C. $2x - y = 2$

    D. $2x + y = -2$

29. A boat travels 30 miles upstream in three hours. It makes the return trip in one and a half hours. What is the speed of the boat in still water?
    (Rigorous) (Skill 3.5)

    A. 10 mph

    B. 15 mph

    C. 20 mph

    D. 30 mph

30. Three less than four times a number is five times the sum of that number and 6. Which equation could be used to solve this problem?
    (Average) (Skill 3.5)

    A. $3 - 4n = 5(n + 6)$

    B. $3 - 4n + 5n = 6$

    C. $4n - 3 = 5n + 6$

    D. $4n - 3 = 5(n + 6)$

31. State the domain of the function $f(x) = \dfrac{3x - 6}{x^2 - 25}$
    (Rigorous) (Skill 3.7)

    A. $x \neq 2$

    B. $x \neq 5, -5$

    C. $x \neq 2, -2$

    D. $x \neq 5$

32. A school band has 200 members. Looking at the pie chart below, determine which statement is true about the band.
    (Easy) (Skill 4A.2)

    A. There are more trumpet players than flute players

    B. There are fifty oboe players in the band

    C. There are forty flute players in the band

    D. One-third of all band members play the trumpet

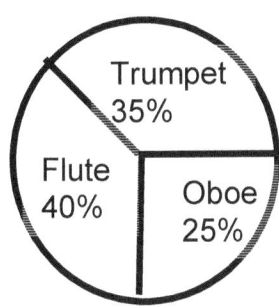

33. On the roll of a standard die, what is the probability of getting a prime number?
    (Easy) (Skill 4A.3)

    A. 1/4

    B. 1/2

    C. 3/4

    D. 1

34. How many ways are there to choose a potato and two green vegetables from a choice of three potatoes and seven green vegetables?
    (Rigorous) (Skill 4A.3)

    A. 126

    B. 63

    C. 21

    D. 252

35. A sack of candy has 3 peppermints, 2 butterscotch drops, and 3 cinnamon drops. One candy is drawn and replaced, then another candy is drawn; what is the probability that both will be butterscotch?
    (Average) (Skill 4A.4)

    A. 1/2

    B. 1/28

    C. 1/4

    D. 1/16

36. Compute the median for the following data set:
    {12, 19, 13, 16, 17, 14}
    (Average) (Skill 4A.7)

    A. 14.5

    B. 15.17

    C. 15

    D. 16

37. Half the students in a class scored 80% on an exam, most of the rest scored 85% except for one student who scored 10%. Which would be the best measure of central tendency for the test scores?
    (Average) (Skill 4A.7)

    A. Mean

    B. Median

    C. Mode

    D. Either the median or the mode because they are equal

38. Half the students in a class scored more than 75 out of a possible 100 points on a math test. Megan scored 75 points. Which of the following statements is true?
    (Easy) (Skill 4A.8)

    A. Megan's score was around the 75th percentile

    B. 25% of the students scored higher than Megan

    C. Megan's score was around the 50th percentile

    D. Megan missed half the questions on the test

39. {1, 4, 7, 10 . . .}

    What is the 40th term in this sequence?
    (Average) (Skill 4B.4)

    A. 43

    B. 121

    C. 118

    D. 120

40. Find the sum of the first one hundred terms in the progression.
(−6, −2, 2 . . .)
(Rigorous) (Skill 4B.4)

   A. 19,200

   B. 19,400

   C. −604

   D. 604

CONSTRUCTED RESPONSE QUESTIONS

1. A man sold two cars for $6,500 each. On the first car he made a profit of 30% and on the second car he lost 30%.

    (a) Did he make a profit overall or did he lose money?
    (b) What was the percentage of his net gain or loss?

2. Two friends live on an east-west street and agree to meet at a park on the same street due east from both of their houses. Nancy leaves home at 8:00 a.m. and drives at 30 mph to the park, which is 25 miles from her house. Susan lives 15 miles away from the park and leaves home at a later time driving at 50 mph. If both of them reach the park at the same time, what time did Susan leave home? Include the following in your solution:

    (a) A diagram showing the locations and known distances in the problem.
    (b) Equations to represent the relationship between distance driven and time for both Nancy and Susan.
    (c) A graph of the above equations. Explain what each element of the graph represents.

3. Find the area of the figure bounded by lines joining the points (0, 0), (0,5), (5,10), (10,5), (5,0), (0,0) in the order given.

    (a) Draw and label the figure on a coordinate plane.
    (b) Explain every step of your reasoning.

Praxis Middle School Mathematics 0069
Post-Test Sample Questions with Rationales

The following represent one way to solve the problems and obtain a correct answer. There are many other mathematically correct ways of determining the correct answer.

1. $7t - 4 \cdot 2t + 3t \cdot 4 \div 2 =$
   (*Average*) (*Skill 1.2*)

   A.   5t

   B.   0

   C.   31t

   D.   18t

Answer: A. 5t
First perform multiplication and division from left to right; 7t −8t + 6t, then add and subtract from left to right.

2. Which statement is an example of the identity axiom of addition?
   (*Easy*) (*Skill 1.3*)

   A. $3 + -3 = 0$

   B. $3x = 3x + 0$

   C. $3 \cdot \frac{1}{3} = 1$

   D. $3 + 2x = 2x + 3$

Answer: B. 3x = 3x + 0
This choice illustrates the identity axiom of addition. Choice A illustrates additive inverse, choice C illustrates the multiplicative inverse, and choice D illustrates the commutative axiom of addition.

3. **Which of the following does not correctly relate an inverse operation?**
    (Average) (Skill 1.4)

    A. $a - b = a + -b$

    B. $a \times b = b \div a$

    C. $\sqrt{a^2} = a$

    D. $a \times \frac{1}{a} = 1$

Answer: B. $a \times b = b \div a$
Choice B is always false. Choices A, C, and D illustrate various properties of inverse relations.

4. **Change $.\overline{63}$ into a fraction in simplest form.**
    (Average) (Skill 1.6)

    A. 63/100

    B. 7/11

    C. 6 3/10

    D. 2/3

Answer: B. 7/11
Let N = .636363…. Then multiplying both sides of the equation by 100 or 10^2 (because there are 2 repeated numbers), we get 100N = 63.636363… Then subtracting the two equations gives 99N = 63 or N = $\frac{63}{99} = \frac{7}{11}$.

5. **Which of the following is an irrational number?**
   (Easy) (Skill 1.7)

   A. .362626262...

   B. $4\frac{1}{3}$

   C. $\sqrt{5}$

   D. $-\sqrt{16}$

Answer: C. $\sqrt{5}$
5 is an irrational number; choices A and B can both be expressed as fractions. Choice D can be simplified to –4, an integer and rational number.

6. $(3.8 \times 10^{17}) \times (.5 \times 10^{-12})$
   (Average) (Skill 1.8)

   A. 19×10^5

   B. 1.9×10^5

   C. 1.9×10^6

   D. 1.9×10^7

Answer: B. 1.9×10^5
Multiply the decimals, and add the exponents.

7. Given even numbers x and y, which could be the LCM of x and y?
   (Average) (Skill 1.11)

   A. $\frac{xy}{2}$

   B. $2xy$

   C. $4xy$

   D. xy

Answer: A. $\frac{xy}{2}$

Although choices B, C and D are common multiples, when both numbers are even, the product can be divided by two to obtain the least common multiple.

8. Simplify: $\dfrac{10}{1+3i}$
   (Rigorous) (Skill 1.16)

   A. $-1.25(1-3i)$

   B. $1.25(1+3i)$

   C. $1+3i$

   D. $1-3i$

Answer: D. $1-3i$

Multiplying numerator and denominator by the conjugate gives
$$\frac{10}{1+3i} \times \frac{1-3i}{1-3i} = \frac{10(1-3i)}{1-9i^2} = \frac{10(1-3i)}{1-9(-1)} = \frac{10(1-3i)}{10} = 1-3i.$$

9. **Evaluate** $3^{1/2}(9^{1/3})$
   (Rigorous) (Skill 1.17)

   A. $27^{5/6}$

   B. $9^{7/12}$

   C. $3^{5/6}$

   D. $3^{6/7}$

Answer: B. $9^{7/12}$

Getting the bases the same gives us $3^{\frac{1}{2}}3^{\frac{2}{3}}$. Adding exponents gives $3^{\frac{7}{6}}$. Then some additional manipulation of exponents produces $3^{\frac{7}{6}} = 3^{\frac{14}{12}} = (3^2)^{\frac{7}{12}} = 9^{\frac{7}{12}}$.

10. **Simplify:** $\sqrt{27} + \sqrt{75}$
    (Rigorous) (Skill 1.17)

    A. $8\sqrt{3}$

    B. 34

    C. $34\sqrt{3}$

    D. $15\sqrt{3}$

Answer: A. $8\sqrt{3}$

Simplifying radicals gives $\sqrt{27} + \sqrt{75} = 3\sqrt{3} + 5\sqrt{3} = 8\sqrt{3}$.

11. $\dfrac{3.5 \times 10^{-10}}{0.7 \times 10^{4}}$

    (Rigorous) (Skill 1.17)

    A. 0.5×10^{6}

    B. 5.0×10^{-6}

    C. 5.0×10^{-14}

    D. 0.5×10^{-14}

Answer: C. 5.0×10^{-14}
Divide the decimals, and subtract the exponents.

12. Solve for x and y:
    x = 3y + 7
    7x + 5y = 23
    (Rigorous) (Skill 1.19)

    A. (−1, 4)

    B. (4, −1)

    C. ($\dfrac{-29}{7}$, $\dfrac{-26}{7}$)

    D. (10, 1)

Answer: B. (4, −1)
Substituting x in the second equation results in 7(3y + 7) + 5y = 23. Solve by distributing and grouping like terms: 26y+49 = 23, 26y = −26, y = −1   Substitute y into the first equation to obtain x.

13. Which of the following is a factor of $k^3 - m^3$?
    (Average) (Skill 1.20)

    A. $k^2 + m^2$

    B. $k + m$

    C. $k^2 - m^2$

    D. $k - m$

Answer: D. k – m
The complete factorization for a difference of cubes is (k – m) (k² + mk + m2).

14. Which of the following is a term in an expansion of the binomial $(x+y)^7$?
    (Rigorous) (Skill 1.20)

    A. $7x^7$

    B. $21x^5 y^2$

    C. $21x^4 y^3$

    D. $7xy^7$

Answer: B. $21x^5 y^2$
Choices A and D are clearly wrong since the first term, x^7, in the expansion will have a coefficient of 1 and the powers of x and y in choice D add up to 8 instead of 7. The coefficient for the term $x^5 y^2$ is given by $\frac{7!}{5!2!} = 21$. The coefficient for the term $x^4 y^3$ is given by $\frac{7!}{4!3!} = 35$. Hence, choice C is the correct choice.

15. **Solve for** x: $18 = 4 + |2x|$
    (Average) (Skill 1.21)

    A. $\{-11, 7\}$

    B. $\{-7, 0, 7\}$

    C. $\{-7, 7\}$

    D. $\{-11, 11\}$

Answer: C. $\{-7, 7\}$
Using the definition of absolute value, two equations are possible: 18 = 4 + 2x or 18 = 4 – 2x. Solving for x gives x = 7 or x = –7.

16. **Determine the area of the shaded region of the trapezoid in terms of** *x* **and** *y*.
    (Rigorous) (Skill 2.2)

    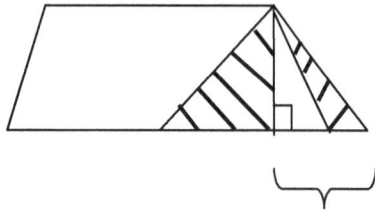

    A. $4xy$

    B. $2xy$

    C. $3x^2 y$

    D.   There is not enough information given

Answer: B. $2xy$
To find the area of the shaded region, find the area of triangle ABC and then subtract the area of triangle DBE. The area of triangle ABC is .5(6x)(y) = 3xy. The area of triangle DBE is .5(2x)(y) = xy. The difference is 2xy.

17. Find the height of a box with surface area of 94 sq. ft. with a width of 3 feet and a depth of 4 feet.
    (Rigorous) (Skill 2.2)

    A. 3 ft.

    B. 4 ft.

    C. 5 ft

    D. 6 ft.

Answer: C. 5 ft.
$94 = 2(3h) + 2(4h) + 2(12)$
$94 = 6h + 8h + 24$
$94 = 14h + 24$
$70 = 14h$
$5 = h$

18. Ginny and Nick head back to their respective colleges after being home for the weekend. They leave their house at the same time and drive for 4 hours. Ginny drives due south at the average rate of 60 miles per hour, and Nick drives due east at the average rate of 60 miles per hour. What is the straight-line distance between them, in miles, at the end of the 4 hours?
(Rigorous) (Skill 2.3)

A. $120\sqrt{2}$

B. 240

C. $240\sqrt{2}$

D. 288

Answer: C. $240\sqrt{2}$
Draw a picture.

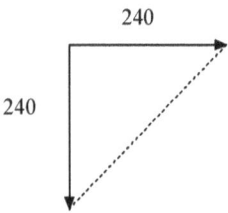

We have a right triangle, so we can use the Pythagorean Theorem to find the distance between the two points.

$$240^2 + 240^2 = c^2$$
$$2(240)^2 = c^2$$
$$240\sqrt{2} = c$$

19. When you begin by assuming the conclusion of a theorem is false, then show that through a sequence of logically correct steps you contradict an accepted fact, this is known as
   (Easy) (Skill 2.6)

   A.   Inductive reasoning

   B.   Direct proof

   C.   Indirect proof

   D.   Exhaustive proof

Answer: C. Indirect proof
By definition this describes the procedure of an indirect proof.

20. Given $l_1 \parallel l_2$ which of the following is true?
   (Average) (Skill 2.6)

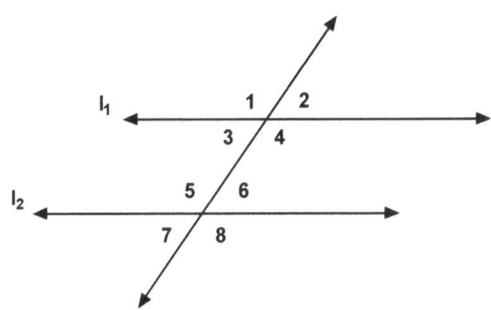

   A.   ∠1 and ∠8 are congruent and alternate interior angles

   B.   ∠2 and ∠3 are congruent and corresponding angles

   C.   ∠3 and ∠4 are adjacent and supplementary angles

   D.   ∠3 and ∠5 are adjacent and supplementary angles

Answer: C. ∠3 and ∠4 are adjacent and supplementary angles
The angles in choice A are exterior. In choice B, the angles are vertical. The angles in choice D are consecutive, not adjacent.

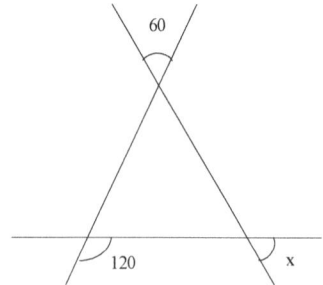

Note: Figure not drawn to scale.

21. In the figure above, what is the value of *x*?
(Rigorous) (Skill 2.7)

    A.   50

    B.   60

    C.   75

    D.   80

Answer: B. 60
The angles within the triangle make up 180°. Opposite angles are equal; therefore, the angle opposite the 60° angle is also 60°. Adjacent angles add to 180° (straight line). Therefore, the angle inside the triangle adjacent to the 120° angle is 60°. The third angle in the triangle would then be 60° (180 – 60 – 60). Since *x* is opposite this third angle, it would also be 60°.

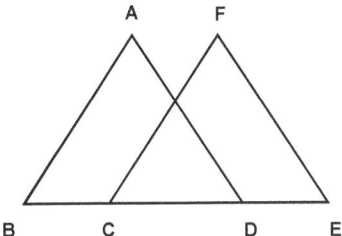

22. **Which postulate could be used to prove △ABD ≅ △CEF, given BC ≅ DE, ∠C ≅ ∠D, and AD ≅ CF?**
    (Average) (Skill 2.7)

    A.   ASA

    B.   SAS

    C.   SAA

    D.   SSS

Answer: B. SAS
To obtain the final side, add CD to both BC and ED.

23. **Given that QO⊥NP and QO=NP, quadrilateral NOPQ can most accurately be described as a**
    (Easy) (Skill 2.8)

    A.   Parallelogram

    B.   Rectangle

    C.   Square

    D.   Rhombus

Answer: C. Square
In an ordinary parallelogram, the diagonals are not perpendicular or equal in length. In a rectangle, the diagonals are not necessarily perpendicular. In a rhombus, the diagonals are not equal in length. In a square, the diagonals are both perpendicular and congruent.

24. What is the slope of any line parallel to the line
    2x + 4y = 4?
    (Rigorous) (Skill 2.12)

    A.   -2

    B.   -1

    C.   -½

    D.   2

Answer: C. -½
The formula for slope is y = mx + b, where m is the slope. Lines that are parallel have the same slope.

$$2x + 4y = 4$$
$$4y = -2x + 4$$
$$y = \frac{-2x}{4} + 1$$
$$y = \frac{-1}{2}x + 1$$

25. Find the midpoint of (2, 5) and (7, –4).
    (Average) (Skill 2.12)

    A.   (9, –1)

    B.   (5, 9)

    C.   (9/2, –1/2)

    D.   (9/2, 1/2)

Answer: D. (9/2, 1/2)
Using the midpoint formula x = (2 + 7)/2 y = (5 + –4)/2

26. **Which set illustrates a function?**
    (Easy) (Skill 3.2)

    A.   { (0,1) (0,2) (0,3) (0,4) }

    B.   {(3, 9) (−3, 9)(4, 16)(− 4, 16)}

    C.   {(1, 2) (2, 3) (3, 4) (1, 4)}

    D.   {(2, 4) (3, 6) (4, 8) (4, 16)}

Answer: B. {(3, 9) (−3, 9) (4, 16) (− 4, 16)}
Each number in the domain can only be matched with one number in the range. A is not a function because 0 is mapped to 4 different numbers in the range. In C, 1 is mapped to two different numbers. In D, 4 is also mapped to two different numbers.

27. **Graph the solution:**
    $|x| + 7 < 13$
    (Rigorous) (Skill 3.3)

    A)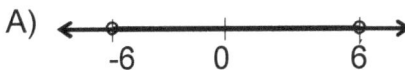

    B)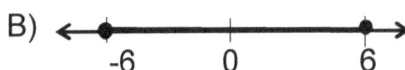

    C)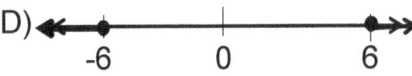

    D)

Answer: A
Solve by adding −7 to each side of the inequality. Since the absolute value of x is less than 6, x must be between −6 and 6. The end points are not included so the circles on the graph are hollow.

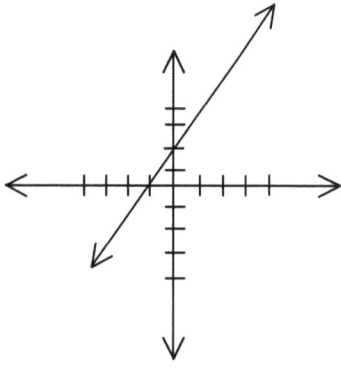

28. What is the equation of the above graph?
(Average) (Skill 3.3)

A. $2x + y = 2$

B. $2x - y = -2$

C. $2x - y = 2$

D. $2x + y = -2$

Answer: B. $2x - y = -2$

By observation, we see that the graph has a y-intercept of 2 and a slope of 2/1 = 2. Therefore its equation is y = mx + b = 2x + 2. Rearranging the terms gives 2x – y = –2.

29. A boat travels 30 miles upstream in three hours. It makes the return trip in one and a half hours. What is the speed of the boat in still water?
    (Rigorous) (Skill 3.5)

    A.   10 mph

    B.   15 mph

    C.   20 mph

    D.   30 mph

Answer: B. 15 mph
Let x = the speed of the boat in still water and c = the speed of the current.

	rate	time	distance
upstream	$x - c$	3	30
downstream	$x + c$	1.5	30

Solve the system:

$3x - 3c = 30$
$1.5x + 1.5c = 30$

30. Three less than four times a number is five times the sum of that number and 6. Which equation could be used to solve this problem?
    (Average) (Skill 3.5)

    A. $3 - 4n = 5(n + 6)$

    B. $3 - 4n + 5n = 6$

    C. $4n - 3 = 5n + 6$

    D. $4n - 3 = 5(n + 6)$

Answer: D. $4n - 3 = 5(n + 6)$
Be sure to enclose the sum of the number and 6 in parentheses.

31. State the domain of the function $f(x) = \dfrac{3x-6}{x^2-25}$

    (Rigorous) (Skill 3.7)

    A. $x \neq 2$

    B. $x \neq 5, -5$

    C. $x \neq 2, -2$

    D. $x \neq 5$

Answer: B. $x \neq 5, -5$

The values of 5 and –5 must be omitted from the domain of all real numbers because if x took on either of those values, the denominator of the fraction would have a value of 0, and therefore the fraction would be undefined.

32. A school band has 200 members. Looking at the pie chart below, determine which statement is true about the band.
    (Easy) (Skill 4A.2)

    A. There are more trumpet players than flute players

    B. There are fifty oboe players in the band

    C. There are forty flute players in the band

    D. One-third of all band members play the trumpet

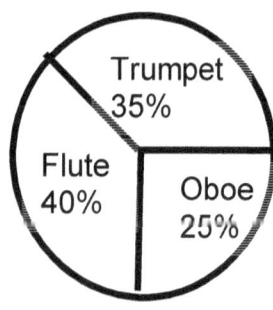

Answer: B. There are fifty oboe players in the band
25% of 200 is 50.

33. **On the roll of a standard die, what is the probability of getting a prime number?**
    (Easy) (Skill 4A.3)

    A.   1/4

    B.   1/2

    C.   3/4

    D.   1

Answer: B. 1/2
Three of the six numbers on the sides of a die (2, 3, and 5) are prime numbers. Hence the probability of getting a prime number is 3/6 = 1/2.

34. **How many ways are there to choose a potato and two green vegetables from a choice of three potatoes and seven green vegetables?**
    (Rigorous) (Skill 4A.3)

    A.   126

    B.   63

    C.   21

    D.   252

Answer: B. 63
There are 3 ways to choose a potato and the number of ways to choose 2 green vegetables from 7 is given by $\frac{7!}{5!2!} = 21$. Hence the total number of choices = 3 x 21 = 63.

35. A sack of candy has 3 peppermints, 2 butterscotch drops, and 3 cinnamon drops. One candy is drawn and replaced, then another candy is drawn; what is the probability that both will be butterscotch?
(Average) (Skill 4A.4)

   A.   1/2

   B.   1/28

   C.   1/4

   D.   1/16

Answer: D. 1/16
With replacement, the probability of obtaining a butterscotch drop on the first draw is 2/8, and the probability of drawing a butterscotch drop on the second draw is also 2/8. Multiply and reduce to lowest terms.

36. Compute the median for the following data set:
    {12, 19, 13, 16, 17, 14}
    (Average) (Skill 4A.7)

   A.   14.5

   B.   15.17

   C.   15

   D.   16

Answer: C. 15
Arrange the data in ascending order: 12, 13, and 14,16,17,19. The median is the middle value in a list with an odd number of entries. When there is an even number of entries, the median is the mean of the two center entries. Here the average of 14 and 16 is 15.

37. **Half the students in a class scored 80% on an exam, most of the rest scored 85% except for one student who scored 10%. Which would be the best measure of central tendency for the test scores?**
   (Average) (Skill 4A.7)

   A.   Mean

   B.   Median

   C.   Mode

   D.   Either the median or the mode because they are equal

Answer: B. Median
In this set of data, the median (see #14) would be the most representative measure of central tendency since the median is independent of extreme values. Because of the 10% outlier, the mean (average) would be disproportionately skewed. In this data set, it is true that the median and the mode (number which occurs most often) are the same, but the median remains the best choice because of its special properties.

38. **Half the students in a class scored more than 75 out of a possible 100 points on a math test. Megan scored 75 points. Which of the following statements is true?**
   (Easy) (Skill 4A.8)

   A.   Megan's score was around the 75th percentile

   B.   25% of the students scored higher than Megan

   C.   Megan's score was around the 50th percentile

   D.   Megan missed half the questions on the test

Answer: C. Megan's score was around the 50th percentile.
Percentiles divide data into 100 equal parts. A person whose score falls in the 50th percentile has outperformed 50 percent of all those who took the test. This does not mean that the score was 50 percent out of 100 nor does it mean that 50 percent of the questions answered were correct. It means that the grade was higher than 50 percent of all those who took the test.

39. {1,4,7,10, ...}

    What is the 40th term in this sequence?
    (Average) (Skill 4B.4)

    A.   43

    B.   121

    C.   118

    D.   120

Answer: C. 118
This is an arithmetic sequence with first term 1 and common difference 3. Using the formula $a_n = a_1 + (n-1)d$, the 40th term = $1 + (40-1)3 = 118$.

40. **Find the sum of the first one hundred terms in the progression.**
    (−6, −2, 2 . . .)
    (Rigorous) (Skill 4B.4)

    A.   19,200

    B.   19,400

    C.   −604

    D.   604

Answer: A. 19,200
To find the 100th term: $t_{100} = -6 + 99(4) = 390$. To find the sum of the first 100 terms: $S = \dfrac{100}{2}(-6 + 390) = 19200$.

SAMPLE ANSWERS TO CONSTRUCTED RESPONSE QUESTIONS

1. **A man sold two cars for $6,500 each. On the first car he made a profit of 30% and on the second car he lost 30%.**

    (a) Did he make a profit overall or did he lose money?
    (b) What was the percentage of his net gain or loss?

Answer

(a) Since both cars were sold for the same price, the first car was cheaper than the second. Since the percentage is the same in both cases, the profit on the first car was smaller than the loss on the second car. Hence, even without calculating the actual amounts, one can conclude that the man lost money overall.

(b) The cost x of the first car can be calculated by setting up the following proportion equation:

$$\frac{130}{100} = \frac{3500}{x}$$

Cross-multiplying and solving for x,

$$130x = 350000$$
$$\Rightarrow x = \frac{350000}{130} = 2692.31$$

The cost y of the second car can be calculated by setting up the following proportion equation:

$$\frac{70}{100} = \frac{3500}{y}$$

Cross-multiplying and solving for y,

$$70y = 350000$$
$$\Rightarrow y = \frac{350000}{70} = 5000$$

Net cost of the two cars = $7692.31
Net sale price for the cars = $7000

Hence, net percentage loss = $\frac{7692.31 - 7000}{7692.31} \times 100 = \frac{692.31}{7692.31} \times 100 = 9\%$

2. Two friends live on an east-west street and agree to meet at a park on the same street due east from both of their houses. Nancy leaves home at 8:00 a.m. and drives at 30 mph to the park, which is 25 miles from her house. Susan lives 15 miles away from the park and leaves home at a later time driving at 50 mph. If both of them reach the park at the same time, what time did Susan leave home? Include the following in your solution:

(a) A diagram showing the locations and known distances in the problem.
(b) Equations to represent the relationship between distance driven and time for both Nancy and Susan.
(c) A graph of the above equations. Explain what each element of the graph represents.

Answer

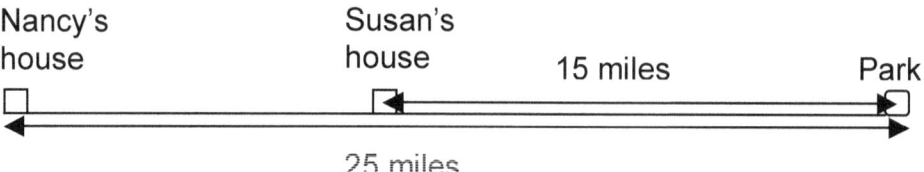

Let d be the distance in miles from Nancy's house, and t the time elapsed in hours after Nancy leaves home. Then $d = 0$ at Nancy's house and $t = 0$ at 8 a.m. Since Nancy drives at a constant speed, the distance traveled by Nancy can be represented by the following linear equation

$d = 30t$

Where the slope of the line is the speed 30 mph.

If Susan leaves home at time t_0 and distance d_0 from Nancy's house, the distance traveled by Susan is given by

$d = 50(t - t_0) + d_0$

Since we know that $d_0 = 10$ miles (distance from Nancy's house to Susan's house), we can write the equation as

$d = 50(t - t_0) + 10$

If both Nancy and Susan reach the park at time t_1, then using equation $d = 30t$, we can find t_1:

$$25 = 30t_1; \; t_1 = \frac{25}{30} = \frac{5}{6} \text{hr}$$

Substituting d = 25 and t = $\frac{5}{6}$ in equation $d = 50(t - t_0) + 10$ we get

$$25 = 50(\frac{5}{6} - t_0) + 10$$

$$\Rightarrow 50(\frac{5}{6} - t_0) = 15$$

$$\Rightarrow \frac{5}{6} - t_0 = \frac{3}{10}$$

$$\Rightarrow t_0 = \frac{5}{6} - \frac{3}{10} = \frac{25-9}{30} = \frac{16}{30} = \frac{8}{15} \text{hr}$$

Thus, Susan left home $\frac{8}{15}$ hr after Nancy at 8:32 a.m.

The above equations are represented graphically as follows:

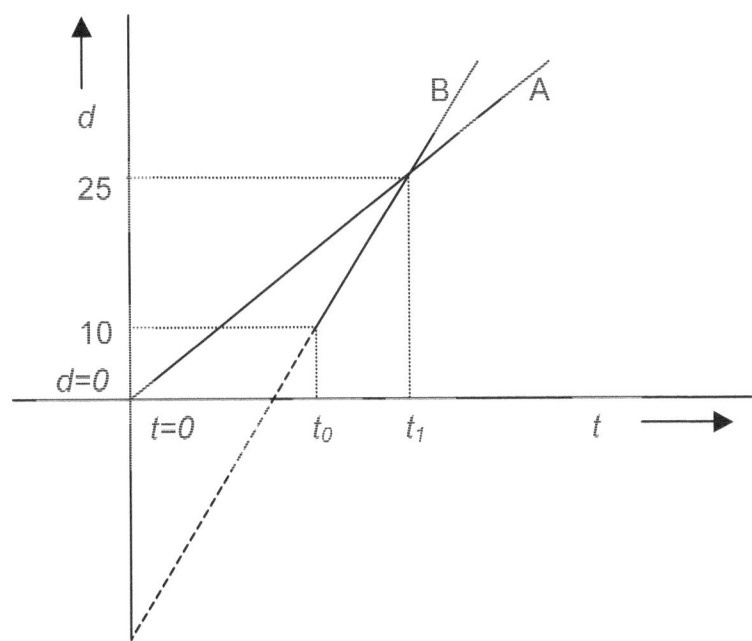

Line A represents the distance driven by Nancy as a function of time. Line B represents the distance driven by Susan. Since both of them drive at constant speeds, the two lines are straight lines. The slopes of the lines represent the speeds at which Nancy and Susan drive. Since Susan drove faster than Nancy, line B has a larger slope. The intersection point of the two lines represents their meeting point at the park and the time at which they reached the park. Note that line B has a negative y-intercept if extended before time t_0. This simply indicates that if Susan had been traveling at the same speed and in the same direction at time t=0 (the time Nancy left home), then she would have started out at a location to the west of Nancy's house.

3. **Find the area of the figure bounded by lines joining the points (0,0), (0,5), (5,10), (10,5), (5,0), (0,0) in the order given.**

   (a) Draw and label the figure on a coordinate plane.
   (b) Explain every step of your reasoning.

Answer

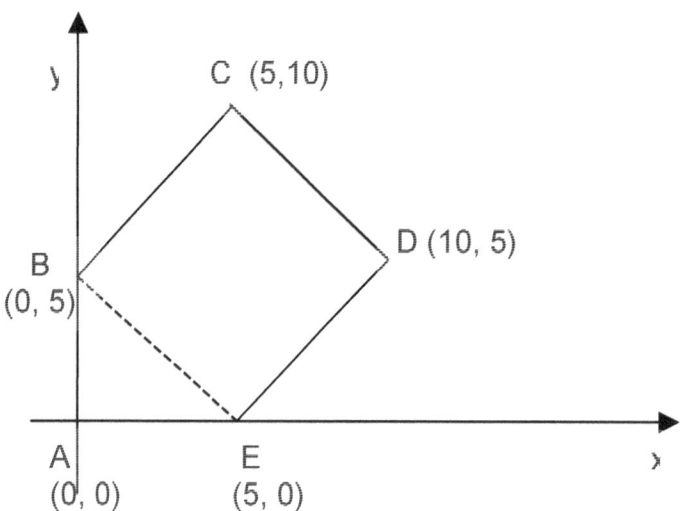

The figure ABCDE is the combination of a right triangle ABE and a square BCDE.

We can demonstrate that BCDE is a square as follows:

Length of line BC = $\sqrt{(0-5)^2 + (5-10)^2} = \sqrt{25+25} = \sqrt{50} = 5\sqrt{2}$

Slope of line BC = $\dfrac{10-5}{5-0} = \dfrac{5}{5} = 1$

Length of line ED = $\sqrt{(5-10)^2 + (0-5)^2} = \sqrt{25+25} = \sqrt{50} = 5\sqrt{2}$

Slope of line ED = $\dfrac{5-0}{10-5} = \dfrac{5}{5} = 1$

Since BC and ED are equal in length and have the same slope, BCDE is a parallelogram. Hence, BE is equal to and parallel to CD.

$$\text{Length of line BE (and CD)} = \sqrt{(0-5)^2 + (5-0)^2} = \sqrt{25+25} = \sqrt{50} = 5\sqrt{2}$$

$$\text{Slope of line BE (and CD)} = \frac{5-0}{0-5} = \frac{5}{-5} = -1$$

BE and CD are equal in length to BC and ED.

Since slope of ED x slope of BE = -1, the two lines are perpendicular to each other. Thus, BCDE is a square with side of length $5\sqrt{2}$.

Therefore, the area of the figure ABCDE = area of ABE + area of BCDE
$= \frac{1}{2} 5 \times 5 + (5\sqrt{2})^2 = 12.5 + 50 = 62.5$ sq. units.

ANSWER KEY

1. A		21. B	
2. B		22. B	
3. B		23. C	
4. B		24. C	
5. C		25. D	
6. B		26. B	
7. A		27. A	
8. D		28. B	
9. B		29. B	
10. A		30. D	
11. C		31. B	
12. B		32. B	
13. D		33. B	
14. B		34. B	
15. C		35. D	
16. B		36. C	
17. C		37. B	
18. C		38. C	
19. C		39. C	
20. C		40. A	

RIGOR TABLE

	Easy 20%	Average 40%	Rigorous 40%
Question	2, 5, 19, 23, 26, 32, 33, 38	1, 3, 4, 6, 7, 13, 15, 20, 22, 25, 28, 30, 35, 36, 37, 39	8, 9, 10, 11, 12, 14, 16, 17, 18, 21, 24, 27, 29, 31, 34, 40

www.ingramcontent.com/pod-product-compliance
Lightning Source LLC
LaVergne TN
LVHW061320060426
835507LV00019B/2241